The ABCs of My Body
(BOOK 1, EXTERNAL)

by Kunle and Ashley Adebiyi

Illustrated by Thomas Petiet

GRACEFUL CARING HANDS, INC.
P.O. Box 31, Saline, Michigan 48176

Copyright © 2020 Kunle and Ashley Adebyi
All rights reserved Printed in the United States

A letter before ABC

The ABCs OF MY BODY (TM) BOOK 1, EXTERNAL) is an opportunity to share with you our experience of teaching our baby, names of her external body parts. We would touch our nose, mouth, eyes and other body parts, naming them and touching our baby's similar parts, repeating the names. This became a fun game that led to the idea of developing a simpler way to make her remember. When we started teaching our baby the alphabet; A is for "Antelope", B is for "Basket"..., we came up with the idea to link all parts of the body to ABCs.

We decided to document this experience in a book and asked Thomas Petiet if he would be interested in illustrating a book with this theme. He showed great enthusiasm and went to work right away. His love of art and personal experience as a father and grandfather are displayed in this book. We are thankful for his encouragement, artistry, arrangement, patience, sacrifice and other contributions to the production of this book.

This book is also intended to teach the alphabets; A, B, C, D…. (a, b, c, d…).It is our hope that this format of teaching the Human Body parts in artistic form and inserting their functions and questions for conversation would stimulate greater interest in children's learning of their unique bodies. We would encourage parents, grandparents, family members, teachers and others to engage with children and use this book as a teaching tool for kids. Children would be enriched by the detailed knowledge of their body parts and functions.

Some letters do not have words for external body parts in English. In such cases, we have indicated fun words such as "O"- Oof!, or "Z"- Zzz. There are equivalent words in other languages such as Ohr (ear in German), Zhou (elbow in Chinese), Visage (face in French), Zahn (tooth in German), Xigai (knee in Chinese).

We have intentionally left out the illustration of external male and female reproductive body parts in the main body of the book. Parents can introduce these words to their children at their discretion and as appropriate.

This book and others to follow would be translated into other languages, such as: Japanese, Chinese, French, German, Russian, Arabic, Spanish, Swahili, Hebrew, Italian Hindi, Bengali and others.

We are grateful to all those who encouraged us and gave us wise counsel and advice. It is our hope that you will use this and other forthcoming books in the series for pleasurable, fun reading, and for imparting knowledge, awareness, and beauty of the uniqueness of our bodies.

This book is dedicated
to the children of the world, who, during these unprecedented times, can still savor the joy of childhood, and to their parents, who can instill in them the ability to love and trust.

A

Arm

Look at me! I swing my **Arm** while I kick the football.
Can you lift up your arm?

Ankle

Yes! The strong **Ankle** keeps my foot stable.
Can you touch your ankle?

B

Belly

My Belly has a button in the middle.
What does my belly do?

B

Back
My **Back** holds my body straight, so I'm not floppy. Can you jump puddles too?

Buttocks
My **Buttocks'** firm muscles move my hips and thighs. They're also good to sit down on.

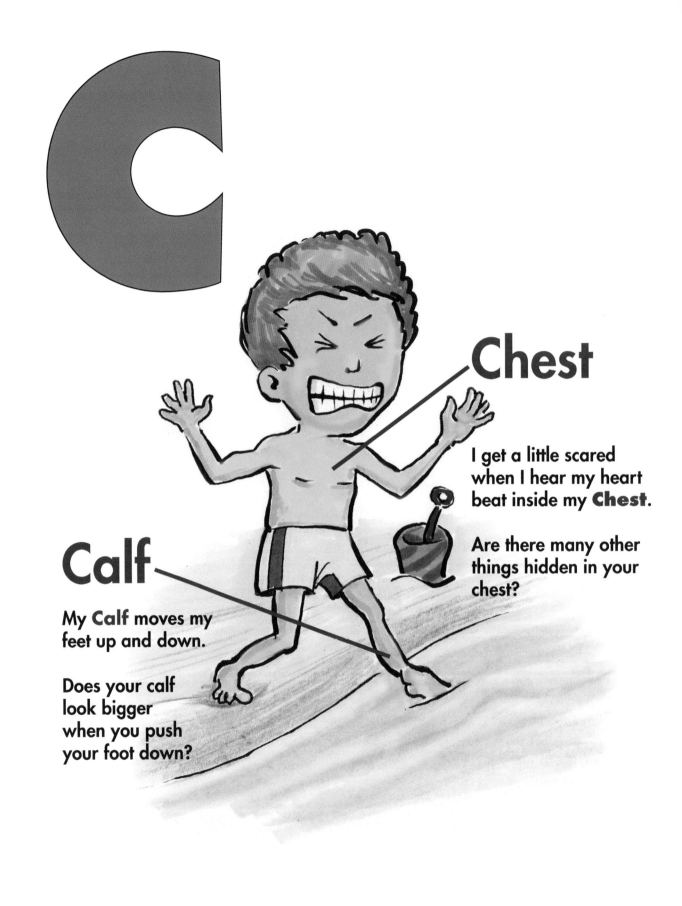

C

Cheek

In addition to my beautiful smile, my **Cheeks** and tongue keep my food between my teeth when I eat.

Chin

My, oh, my. My **Chin** protects my mouth and is a strong part of my jaw. It's a strong feature of my face.

D

Dimples

Dimples are a part of my smile. Where do your dimples come from?

E

Eyebrows

My **Eyebrows** can show my feelings.

Eyelashes

What do my **Eyelashes** do? They protect my eyes from dirt and dust. They're attached to my eyelids that keep my eyes moist when I blink.

Let's see you blink.

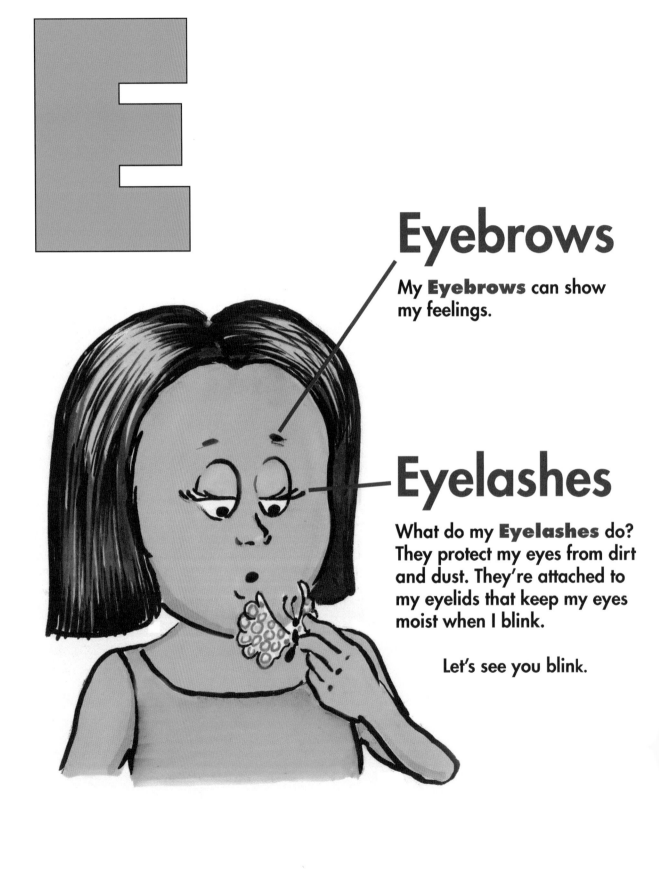

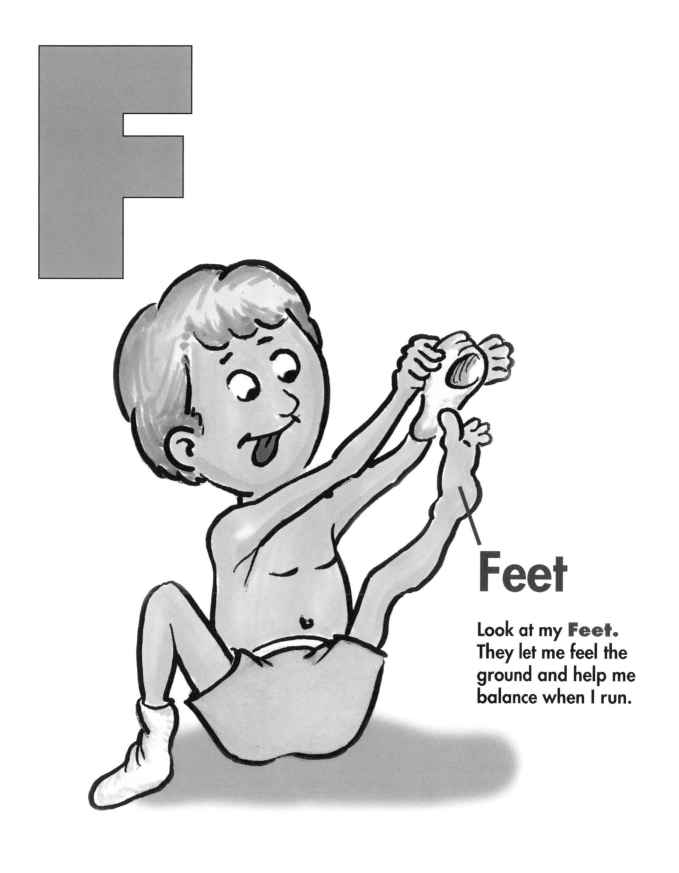

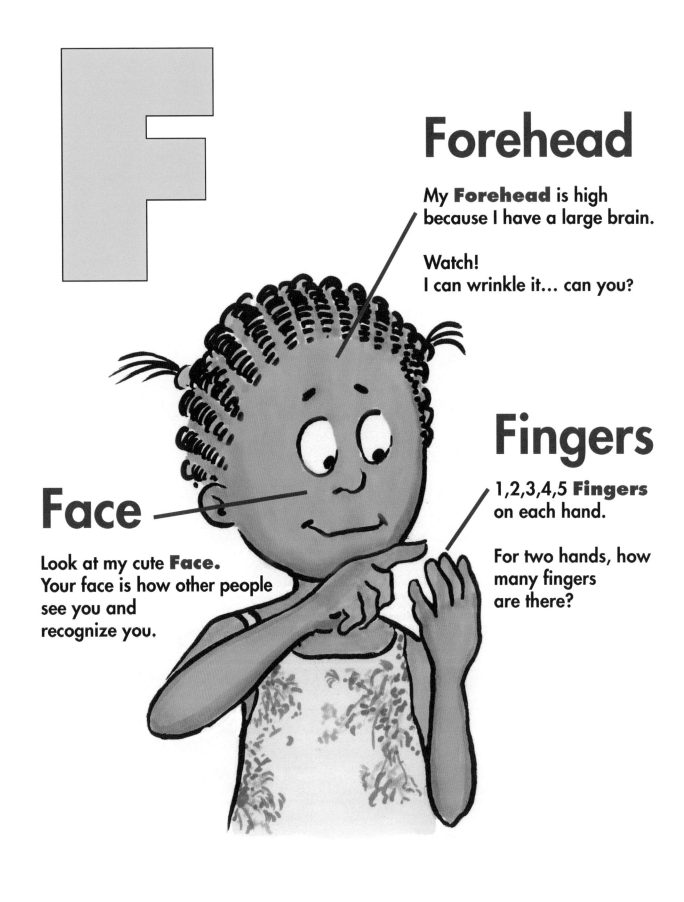

Groin

My Groin is the area between my legs and has strong muscles for my legs to move side to side

H

Head

My **Head** holds my eyes, ears, nose, tongue and even the brain. You should always take good care of it.

Hand

Look at those strong **Hands.** Could you pick anything up without them?

Heels

Your **Heels** are below your ankles. When you walk, your heel comes down first

H

Hair

Hair can protect my skull and help keep my body temperature steady.

It's also pretty, isn't it?

Hip

My **Hips** connect my legs to my backbone. When I dance, I can move my hips up and down.

Index Finger

My **Index Finger** is the first one on my hand and the one I point with.

Do you point with it too?

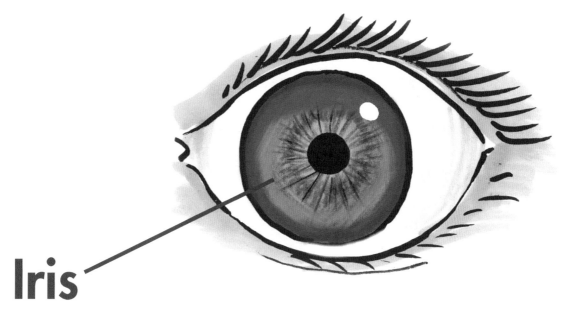

Iris

What is the color of my eye? The **Iris** is the colored part and it opens and closes to control light coming in.

J

Jaw

My Jaw holds my teeth. I exercise my jaw by eating.

Do you like to exercise that way too?

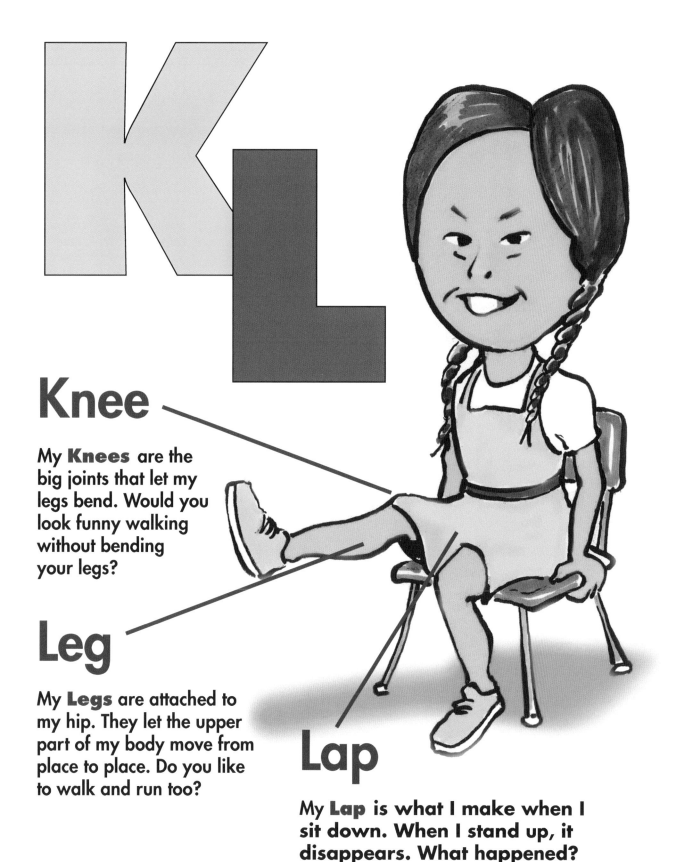

K L

Knee

My **Knees** are the big joints that let my legs bend. Would you look funny walking without bending your legs?

Leg

My **Legs** are attached to my hip. They let the upper part of my body move from place to place. Do you like to walk and run too?

Lap

My **Lap** is what I make when I sit down. When I stand up, it disappears. What happened?

K

Knuckles

My **Knuckles** let me move my fingers to better grab things. Without them, my fingers would break.

L

Lips

La, la, la! I can make sounds with my **Lips.** They help me speak and blow air into a balloon. My lips also keep unwanted things out of my mouth.

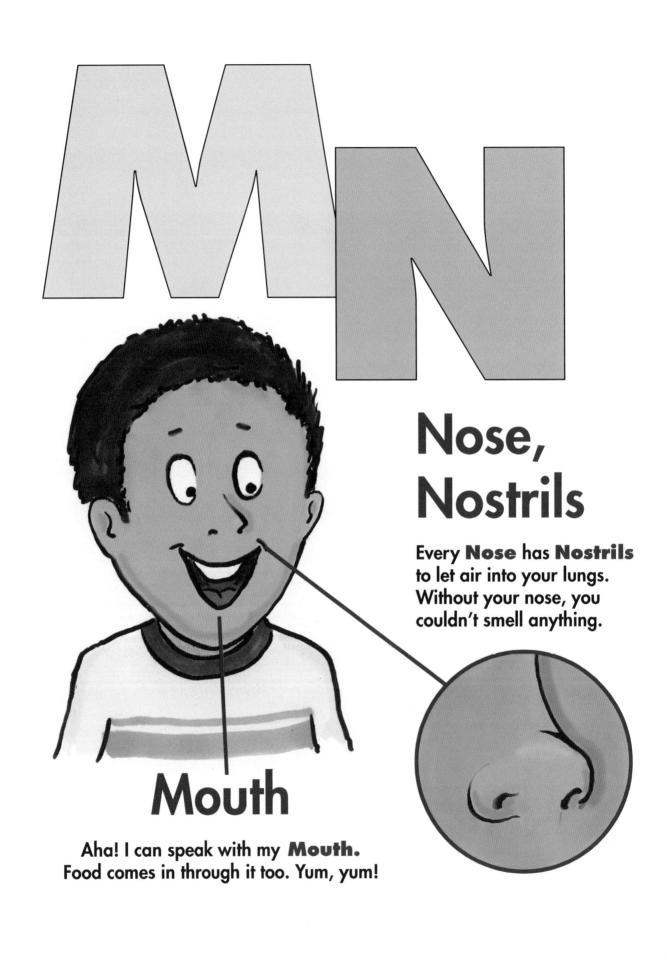

M N

Nose, Nostrils

Every **Nose** has **Nostrils** to let air into your lungs. Without your nose, you couldn't smell anything.

Mouth

Aha! I can speak with my **Mouth.** Food comes in through it too. Yum, yum!

M

Muscle

Can you see this strong **Muscle**?
I have them all over my body to make my bones move.

N

Navel

What's this Navel on my belly? You have one too. It's just the spot where you were connected to your mother before you were born.

N

Neck

Look at my long **Neck.** It attaches my head to my body like a powerful cable. It holds nerves from my brain and can even turn so I can see all around.

What's that up there?

Oof!*

OOF is what I say when I lift too much.

* We're just being silly here, because there really isn't a body part you can see that starts with an O in English. You'll see some in a later book on internal organs. Still, in German there is an O: Ohr (Ear).

P

Patella

You may not think about it much, but the **Patella** is a bone that lies on top of the knee to protect the joint.

Without it, kneeling down would hurt a lot.

P

Pinky Finger

My **Pinky Finger** is the smallest one on my hand.

Do you know that it provides half the strength of the whole hand?

P

Pant!*

Pant! is what I do when I get out of breath from running.

* Pant is not really a body part, but we added it because it's something we all do when we get enough exercise to keep our bodies healthy.

Quiet!

Q is a letter we'll have to wait for when we read "The ABCs of the Muscles" to learn about things like the Quadratus Lumborum.

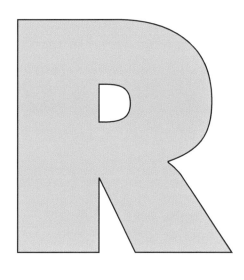

Ribs

You can see my **Ribs** when I bend. They protect all the internal parts of my body, and are attached to the back bone.

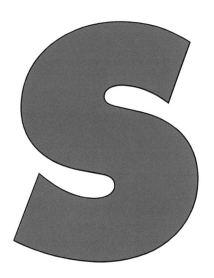

Shoulder

Big hug! My brother likes to lean on my **Shoulder.**

S

Skin

My **Skin** is like clothing for all the parts inside my body. It has to protect me, and sometimes it gets hurt.

Thanks, Mom, for putting lotion on my sunburn!

T

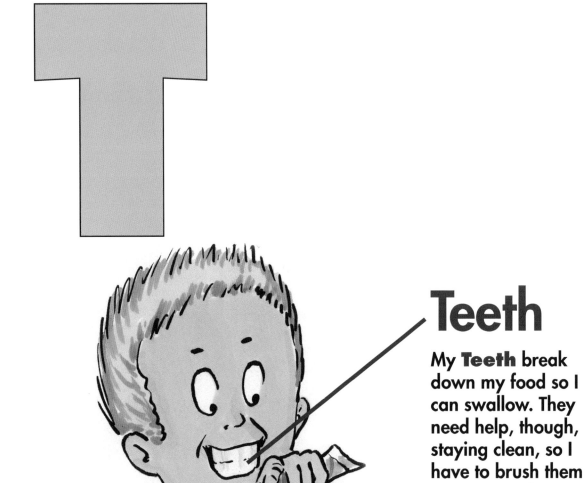

Teeth

My Teeth break down my food so I can swallow. They need help, though, staying clean, so I have to brush them morning and evening.

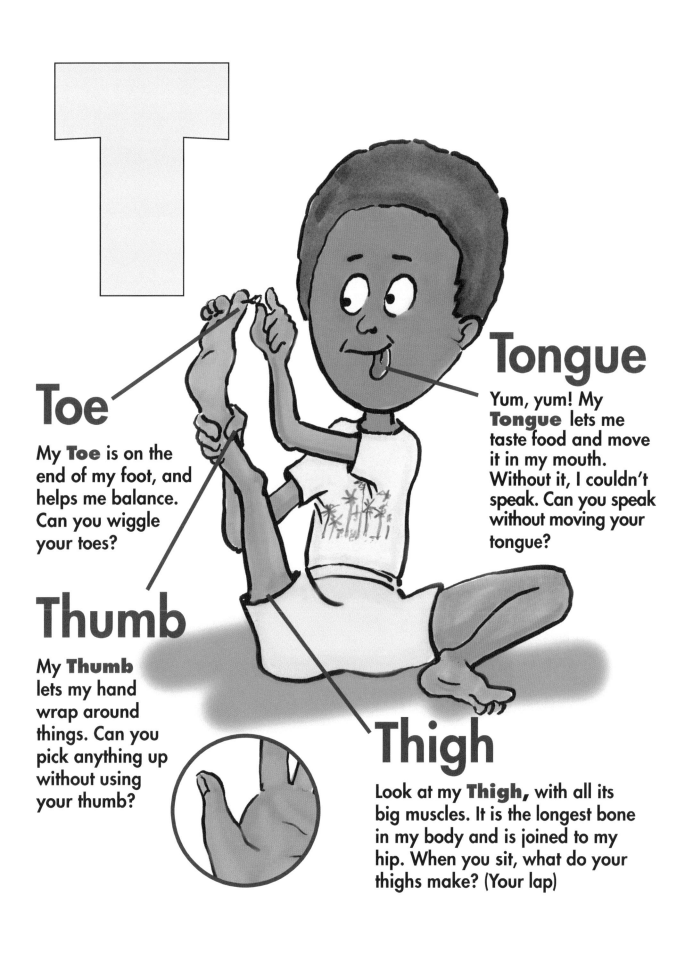

U

Umm...*

Umm is what we sometimes say when we can't quite think of something.

Have you said it too?

* Let's learn a little Japanese, because we don't have a visible body part that starts with U in English. In Japan, the Arm is called Ude.

U

Underwear*

It's not a part of my body, but **Underwear** is always a part of what I wear. It helps keep my clothes clean.

It's usually white, but sometimes it's colored. What color is yours?

* Like Underwear, there are parts of your body that aren't usually seen. You can see one of them if you look in the back of your mouth. It hangs down and is involved in swallowing. It's called the Uvula.

V

Voice*

You can hear me when I use my **Voice.** You can't see it but you can hear it. It's one of the most important parts of the body because it's how we know what other people think.

What was that you said?

* Let us learn a little of the French language. The face is called the Visage in France.

W

Wrist

My **Wrist** lets my hand move up and down and sideways. I also like to put bracelets on it.

Do you have things you put on your wrist?

Waist

My **Waist** is between my chest and my hips.

Do you put a belt around your waist for any reason?

Xx Yy

Yawn*

When I'm sleepy, I Yawn.

Parents yawn.

Many animals yawn.

Do you?

* In English, there are no visible body parts that start with these letters. But let's try other languages. In China, the Chest is called the Xiongbu, and the Knee is the Xigai. In Japan, the Finger is called the Yubi.

Zzz*

In cartoons like this one, zzz is the sound of snoring when you snooze.

Do you snore?

Do your parents snore?

* In English, there are no visible body parts that start with this letter. But in China, they call the Elbow the Zhou. In Germany, a tooth is a Zahn.

About the Authors and Illustrator

Kunle Adebiyi is a father and works as a Clinical Pharmacist at Michigan Medicine (formerly The University of Michigan Health System), Ann Arbor Michigan. He began his working career as a high school science teacher, and later worked for Ciba-Geigy (now Novartis Pharmaceutical). More recently, he co-founded and was the director of a home health care company, serving Washtenaw and Wayne counties in Michigan.

Ashley Renee Adebiyi is a devoted wife and mother. She holds a Masters of Arts degree in Clinical Mental Health Counseling. She has worked as a health care giver in Wayne and Washtenaw counties, and later, as an administrative assistant at a home health care company in southeast Michigan.

Thomas Petiet is a father who has enjoyed a long career as a commercial artist and illustrator. As Concept Studio, founded in 1971, Mr. Petiet has produced brochures, space ads, catalogs and illustrations. He was the art director for the Creative Activities book series for children. He has been the Managing Director of The Comic Opera Guild since 1973, producing over 100 shows and numerous recordings

Upcoming books in the ABCs of My Body series

"The ABCs of My Body Internal"

Digestive system

Muscular system

Nervous system

Skeletal; bones

cartilage

Head/Brain

spinal cord

Respiratory System

Heart and Internal organs

Blood & blood vessels

Enzymes

Hormone and

Humoral system.

Printed in Great Britain
by Amazon